TABLE OF CONTENTS

Introduction — 4

Chapter 1 What Is Cryptology? — 6

Chapter 2 Cryptologists Through the Ages — 10

Chapter 3 How Technology Changed Cryptology — 16

Chapter 4 Cryptologists in the Digital Age — 22

Conclusion — 29

Crack It! The Caesar Shift — 29

Glossary — 30

Learn More — 31

Index — 32

INTRODUCTION

It was 1917, and World War I (1914–1918) was raging in Europe. British cryptologists pored over the intercepted messages of German soldiers. The cryptologists worked together to crack the German code system.

In January, the British intercepted a telegram on its way to Mexico. It came from German foreign secretary Arthur Zimmermann. After all their hard work, the British cryptologists were able to decode the message. The secret it revealed was staggering. Zimmermann was offering to help Mexico invade the southern United States!

Until then, the United States had not been involved in World War I. But learning of Germany's offer to Mexico helped convince the United States to join the fight. Cryptologists had changed the course of history.

Arthur Zimmermann

The British cryptologists who intercepted the Zimmerman Telegram worked in the Old Admiralty Building in London, England.

CHAPTER 1

WHAT IS CRYPTOLOGY?

Cryptology is the science of secret communication. People called cryptologists have studied and practiced cryptology for thousands of years.

There are two types of cryptologists: cryptographers and cryptanalysts. Cryptographers encode or encipher a message to hide its meaning. Cryptanalysts decode or decipher a message's hidden meaning.

Cryptology was once a mysterious profession. Code breakers worked in secret, often for governments. However, the invention of the computer changed cryptology forever. Modern cryptologists keep emails, texts, and online purchases secure. In the digital age, professional cryptology has become an important and popular career.

Modern cryptology often involves computer science and math.

CRYPTOLOGY
PROFESSIONAL CRYPTOLOGISTS

RACHAEL L. THOMAS

Lerner Publications ◆ Minneapolis

ALTERNATOR BOOKS™

Copyright © 2022 by Lerner Publishing Group, Inc.

All rights reserved. International copyright secured. No part of this book may be reproduced, stored in a retrieval system, or transmitted in any form or by any means—electronic, mechanical, photocopying, recording, or otherwise—without the prior written permission of Lerner Publishing Group, Inc., except for the inclusion of brief quotations in an acknowledged review.

Lerner Publications Company
An imprint of Lerner Publishing Group, Inc.
241 First Avenue North
Minneapolis, MN 55401 USA

For reading levels and more information, look up this title at www.lernerbooks.com.

Main body text set in Aptifer Sans LT Pro
Typeface provided by Linotype

The images in this book are used with the permission of: © DutchScenery/Getty Images, p. 3; © Library of Congress, pp. 4, 17; © Hulton Archive/Getty Images, p. 5; © metamorworks/Getty Images, pp. 6–7; © Anatolii Mazhora/Shutterstock Images, pp. 8–9; © PaoloGaetano/Getty Images, p. 10; © HansFree/Shutterstock Images, p. 11; © Wikimedia Commons, p. 12; © Culture Club/Getty Images, p. 13; © Szekeres Szabolcs/Shutterstock Images, p. 14; © Nastasic/Getty Images, p. 15; © Christopher Wood/Shutterstock Images, p. 16; © National Archives and Records Administration, pp. 18–19; © Lenscap Photography/Shutterstock Images, p. 20; © National Security Agency, p. 21; © Science & Society Picture Library/Getty Images, p. 22; © Evening Standard/Getty Images, p. 23; © Rawpixel.com/Shutterstock Images, pp. 24–25; © peshkov/Getty Images, p. 26; © Greg Mathieson/Mai/Getty Images, p. 27; © gorodenkoff/Getty Images, p. 28.

Cover Photo: © Flickr/Manuel J. Prieto

Design Elements: © AF-studio/Getty Images; © 4khz/Getty Images; © non-exclusive/Getty Images

Library of Congress Cataloging-in-Publication Data

Names: Thomas, Rachael L., author.
Title: Professional cryptologists / Rachael L. Thomas.
Description: Minneapolis, MN : Lerner Publications Company, an imprint of Lerner Publishing Group, Inc., [2022] | Series: Cryptology. Alternator books | Includes bibliographical references and index. | Audience: Ages 8–12. | Audience: Grades 4–6. | Summary: "Read about the historical code makers and breakers who protected their countries by coding and decoding messages for their governments, and the modern cryptologists who keep emails, texts, and online purchases secure"— Provided by publisher.
Identifiers: LCCN 2020019901 (print) | LCCN 2020019902 (ebook) | ISBN 9781728404585 (library binding) | ISBN 9781728417974 (ebook)
Subjects: LCSH: Cryptographers—Juvenile literature. | Cryptography—Juvenile literature.
Classification: LCC QA268 .T475 2021 (print) | LCC QA268 (ebook) | DDC 652/.8—dc23

LC record available at https://lccn.loc.gov/2020019901
LC ebook record available at https://lccn.loc.gov/2020019902

Manufactured in the United States of America
1-48516-49030-12/3/2020

CIPHERS, CODES, AND KEYS

Ciphers, codes, and keys are the building blocks of cryptology. Ciphers change individual letters to make words unreadable. In a cipher, the word "animal" could become "cpkocn" or "gcengx."

A code affects entire words or phrases. For example, aircraft pilots sometimes say the code words "Roger that." This means "information received."

A cryptographer's work is challenging. A code or cipher must be complex, or it will be cracked. But it must also be possible for allies to easily decrypt. So, cryptologists create keys to help translate specific codes or ciphers.

A key can take many forms. Some are books. Others are special gadgets! No matter what form they take, keys are used to decode the hidden meaning of a message.

A hidden message is called a cryptogram. A cryptogram can be made using a code, a cipher, or a combination of both.

CHAPTER 2
CRYPTOLOGISTS THROUGH THE AGES

Cryptology evolved from written language. The invention of writing meant humans could send long-distance messages. But these messages could easily fall into the wrong hands! The need for secret messages gave cryptology an official purpose.

Roman leader Julius Caesar was one of the first people to put cryptology into action. During a military campaign in the 50s BCE, Caesar wanted to share important news with his armies and generals. So, he used a substitution cipher to encode his messages. Caesar's cipher system switched each letter of a word with the letter that came three positions later in the alphabet.

Julius Caesar

CONCLUSION

The role of the cryptologist has transformed. Cryptologists no longer simply make and break codes. They also work to secure the information of billions of people!

Parts of the job remain the same, however. The US military still hires professional cryptologists to create codes and ciphers for secure communication. Cryptologists also examine secret messages to keep us safe from threats.

Today, cryptology is more popular than ever. Many large universities offer courses in cryptologic studies. In the age of digital information, codes and ciphers will continue to feature in our daily lives . . . whether we see them or not!

Crack It! The Caesar Shift

Use the Caesar Shift to create your own ciphered messages!

A	B	C	D	E	F	G	H	I	J	K	L	M	N	O	P	Q	R	S	T	U	V	W	X	Y	Z
D	E	F	G	H	I	J	K	L	M	N	O	P	Q	R	S	T	U	V	W	X	Y	Z	A	B	C

SEND HELP
VHQG KHOS

GLOSSARY

algorithm: a set of steps used to solve a mathematical problem or complete a computer operation

block cipher: a way of encrypting text in which an algorithm is applied to blocks of data rather than individual bits

cybercrime: criminal activity in which a computer is used to illegally access, send, or change data

data: information in digital form

decipher: to reveal the meaning of a ciphered message

decode: to reveal the meaning of a coded message

decrypt: to find a message's hidden meaning

diplomatic: relating to the work of negotiating between different nations

embassy: the building where an ambassador lives and works

encipher: to hide the meaning of a message using a cipher

encode: to hide the meaning of a message using a code

encrypt: to alter a message to hide its meaning. Once encrypted, a hidden message is called an encryption. The process of encrypting a message is also called encryption.

hacker: someone who illegally gains access to a computer system to steal information or cause damage

instantaneously: happening quickly

intercept: to stop something before it arrives somewhere. Something that has been stopped in this manner has been intercepted.

key: the tool or resource that helps a person decode or decipher a hidden message

recipient: someone who receives something

replica: an exact or close copy of something

statistics: a type of mathematics that deals with the collection and analysis of data

The first known written language was cuneiform. It originated in the Middle East around 3000 BCE.

So, the letters *a*, *b*, and *c* became *d*, *e*, and *f*, and so on. This simple cipher became known as the Caesar Shift. Today, substitution ciphers are a fundamental part of cryptology.

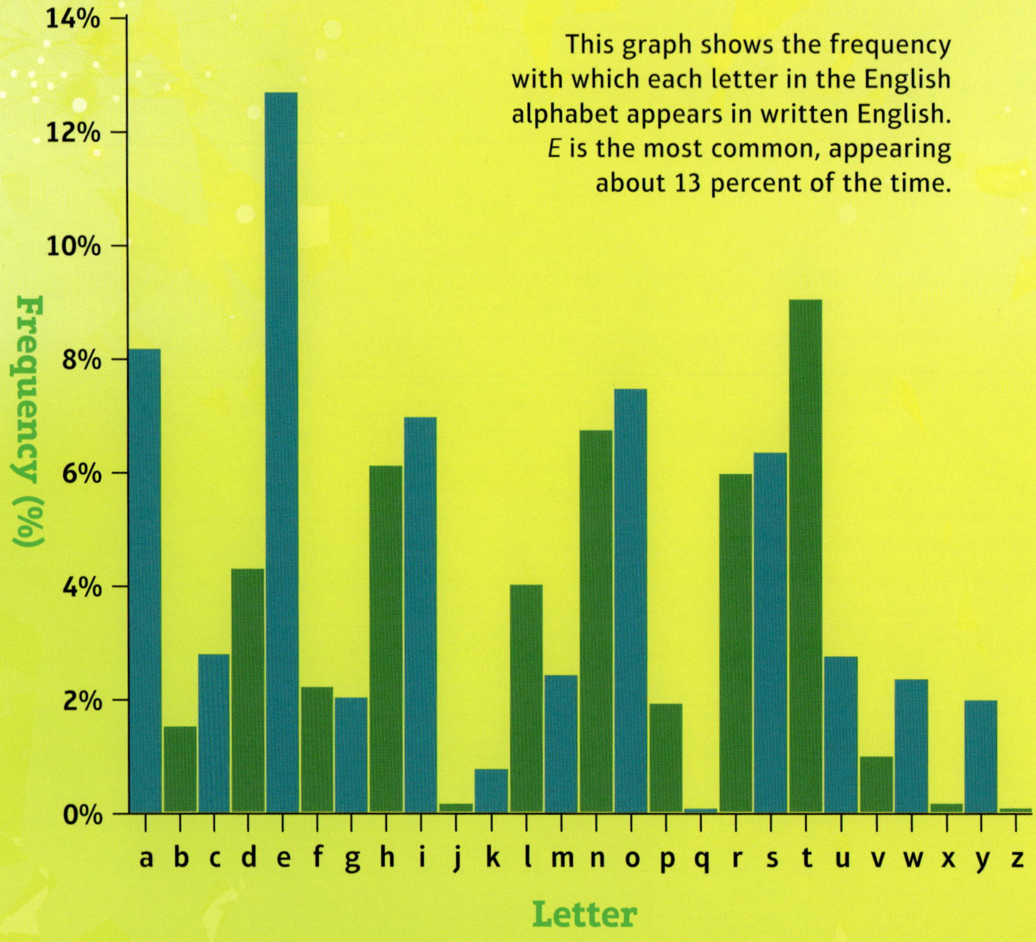

This graph shows the frequency with which each letter in the English alphabet appears in written English. *E* is the most common, appearing about 13 percent of the time.

STEAM Spotlight—Math

Every language has rules and patterns. Frequency analysis uses statistics to spot these patterns in cryptograms. For example, the most common three-letter word in English is "the." By looking at the most frequent group of three letters in a cipher text, you may be able to identify the cipher for the word "the." Knowing this, you can insert *t*'s, *h*'s, and *e*'s into other parts of the cryptogram!

LEARN MORE

Bletchley Park: Enigma
https://bletchleypark.org.uk/our-story/the-challenge/enigma

Central Intelligence Agency: Spy Kids—Break the Code
https://www.cia.gov/kids-page/games/games_code.html

Daigneau, Jean. *Code Cracking for Kids: Secret Communications Throughout History, with 21 Codes and Ciphers.* Chicago: Chicago Review Press, 2020.

Peterson, David J. *Create Your Own Secret Language: Invent Codes, Ciphers, Hidden Messages, and More: A Beginner's Guide.* New York: Odd Dot, 2020.

Schwartz, Ella. *Can You Crack the Code?* New York: Bloomsbury, 2019.

Thomas, Rachael L. *Classic Codes and Ciphers.* Minneapolis: Lerner Publications, 2022.

INDEX

Advanced Encryption System (AES), 24
al-Kindi, 13
al-Ma'mun, 13

black chambers, 14–15
block ciphers, 23

Caesar, Julius, 10
Caesar Shift, 11, 29
Caesar Shift project, 29
codebooks, 19
computers, 6, 22, 24, 26
cryptanalysts, 6, 13, 26
cryptograms, 8, 12, 15
cryptographers, 6, 8, 26
cryptographic algorithms, 24
cryptologic studies, 6, 13, 23, 29
cuneiform, 11
cybercrime, 26

data encryption, 22, 24, 26
Data Encryption Standard (DES), 22–24

Enigma cipher system, 20–21
Enigma machines, 20–21

Feistel, Horst, 23
France, 21
frequency analysis, 12–13

Germany, 4, 19–20
Great Britain, 4–5, 19, 21, 23

International Business Machines (IBM), 22–23

keys, 8, 16, 19, 24, 26

Lucifer algorithm, 22–23

Maryland, 27
Mexico, 4
Middle East, 11
Morse, Samuel, 17
Morse Code, 16–17, 19

National Bureau of Standards (NBS), 22
National Security Agency (NSA), 26–27

Rejewski, Marian, 21

substitution cipher, 10–11, 29

telegraphs, 16–17, 19

United States, 4, 22, 29

Vienna, Austria, 14–15
von Kaunitz, Wenzel Anton, 15

World War I, 4, 19
World War II, 20

Zimmermann, Arthur, 4
Zimmermann Telegram, 19

CRYPTO SPOTLIGHT

Samuel Morse was a co-inventor of the telegraph and the inventor of Morse Code. The code assigns a set of dots and dashes to each letter in the English alphabet. For example, the letter *a* is a dot followed by a dash. To send a telegram message, people tapped out words by combining dashes and dots, or long and short taps.

Samuel Morse

WESTERN UNION TELEGRAM

NEWCOMB CARLTON, PRESIDENT

via Galveston

LEGATION

ICO CITY

2	13401	8501	115	3528	416	17214	64
222	21560	10247	11518	23677	13605		34
05	11311	10392	10371	0302	21290		5161
504	11269	18276	18101	0317	0228		1769
200	19452	21589	67893	5569	13918		895
5	4458	5905	17166	13851	4458	17149	
224	6929	14991	7382	15857	67893		1421
53	67893	5870	5454	16102	15217		22801
388	7446	23638	18222	6719	14331		1502
52	22096	21604	4797	9497	22464		20855
40	22260	5905	13347	20420	39689		1373
	18507	52262	1340	22049	13339		11265
14	4178	6992	8784	7632	7357	6926	

18

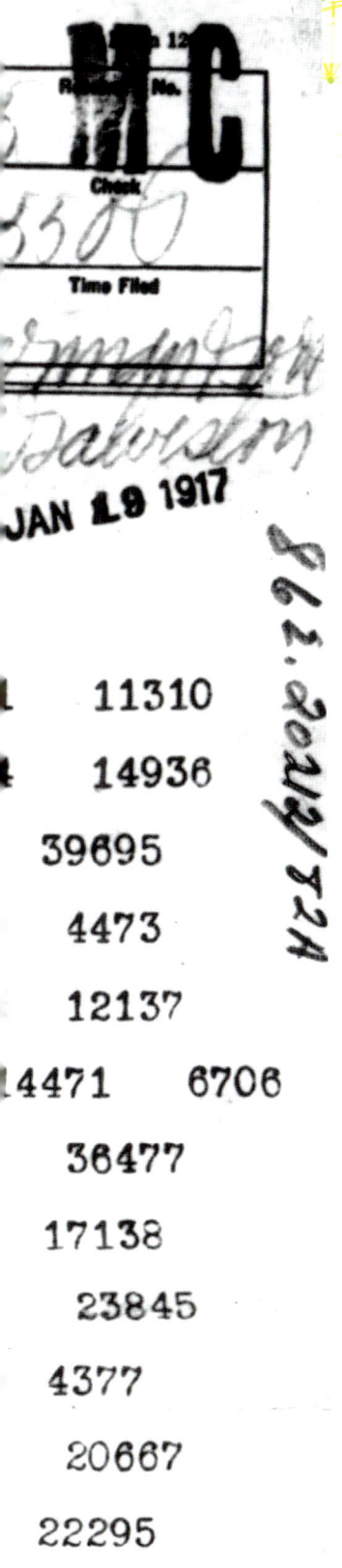

In 1914, World War I broke out across Europe. At the time, people used telegraphs to send messages quickly. But telegrams could be intercepted, and Morse Code was public knowledge. So, some countries used codes and ciphers to further encrypt telegrams.

Wartime cryptologists created keys called codebooks to crack these top-secret codes and ciphers. The books were handed out to trusted allies. Rival nations constantly looked for enemy codebooks to crack secret communications.

In 1914, the British navy found a German codebook in a sunken safe. After studying the book for months, British cryptologists used it to decode Germany's secret wartime messages.

The most famous message was the Zimmermann Telegram. It showed strings of numbers instead of words. British cryptologists used the stolen codebook to decrypt the telegram.

The Zimmermann Telegram

Enigma machines had 158 quintillion possible scrambler settings.

During World War II (1939–1945), countries developed their own secret code or cipher systems. Germany's cipher system was transmitted using devices called Enigma machines. Enigma machines used scrambler disks to create complex ciphers. The scrambler disks constantly rotated. This altered the cipher system even as a message was being written!

To send a message, the sender and recipient first synchronized their machines. The sender typed a message, which their own Enigma machine enciphered and sent to the recipient via radio. The recipient typed the cipher text into their Enigma machine to be decrypted. At midnight, the operators reset their Enigma machines' scrambler disks.

CRYPTO SPOTLIGHT

Cracking Enigma took many people. One important group was a team of three Polish mathematicians. This team analyzed the Enigma cipher system and built a replica Enigma machine. In 1939, they shared what they'd learned with Britain and France. Two years later, Enigma was successfully cracked.

Marian Rejewski was one of the Polish mathematicians who helped crack Enigma.

CRYPTOLOGISTS IN THE DIGITAL AGE

Computers and the internet changed the role of the cryptologist once again. The first personal computer was developed in 1974. With its invention, developers had to consider how digital information could be kept secure.

That year, the National Bureau of Standards (NBS) asked cryptologists to develop an algorithm that might be used to encrypt and protect data. The company International Business Machines (IBM) submitted an algorithm called Lucifer. This algorithm became the Data Encryption Standard, or DES. The DES was the standard for data protection all over the world.

IBM released its Series 1 computer in 1976, one year before the DES was approved.

CRYPTO SPOTLIGHT

IBM employee Horst Feistel designed the Lucifer algorithm that later became the DES. Feistel was one of the first cryptologists to research the design and theory of block ciphers. Block ciphers were a prominent feature of the DES algorithm design.

The Lucifer algorithm was originally designed for Lloyds Bank in London.

At the time, the DES was the strongest cryptographic algorithm available. It protected the top-secret digital information of governments, banks, and businesses.

Cryptographic algorithms like the DES use mathematical codes to encrypt and decrypt data. The secret to this encryption is the key. Algorithms such as the DES create billions upon billions of possible keys. The probability of finding the correct key is very low! So, the encryption becomes almost impossible to crack.

In 2000, the DES was replaced by the Advanced Encryption System, or AES. The AES algorithm increased the key length of the encryption. This created even more possible key options, strengthening the encryption.

As computers become more powerful, cryptographic algorithms must be strengthened and updated.

One study estimates that a hacker attack occurs every 39 seconds somewhere in the world.

The internet has brought new challenges to cryptologists. Cybercrime is a developing area of law. Every day, hackers try to break into secure online networks. Sometimes, they steal private information. Organizations like the National Security Agency (NSA) now have cybersecurity departments to monitor and prevent online criminal activity. The NSA hires cryptologists to strengthen encryption defenses against hackers.

STEAM Spotlight—Technology

Digital data encryption is similar to other forms of cryptography and cryptanalysis. When you send a message using a computer, the text is encrypted and becomes ciphertext. The encrypted message data then travels to another computer, where it must be decrypted. Devices use digital keys in the form of complex algorithms to decrypt messages.

The NSA's headquarters in Fort Meade, Maryland

Military cryptologists have many duties, including deciphering messages written in other languages.

Al-Kindi worked at a research institute established by Caliph al-Ma'mun (*center, with sword*). There, al-Kindi decrypted texts from other nations.

By the Middle Ages, ciphers had become a common method for hiding secret messages. But they were not always secure. Cryptanalysts were developing ways to crack ciphers using linguistics and statistics.

Around 850 CE, Arab philosopher al-Kindi published the world's first academic book about cryptology. In the book, al-Kindi identified frequency analysis as a tool to break ciphers. Frequency analysis made ciphers easier to crack.

In the 1700s, hundreds of letters went through Vienna's black chamber each day.

By the 1700s, there was increasing demand for cryptologists in Europe. Some worked at government cryptology departments called "black chambers." The most famous European black chamber was in Vienna, Austria.

This office intercepted all mail addressed to foreign embassies. Staff hoped to discover sensitive diplomatic information. They opened the letters, wrote out exact copies, and returned the letters to the mail. Cryptologists studied the copied letters for possible cryptograms to crack. Black chambers sold the secrets they learned to other European countries.

The role of the cryptologist was changing. But the codes and ciphers of the 1700s weren't much different from those used earlier in history. In the next centuries, new technology would change cryptology.

Austrian Chancellor Wenzel Anton von Kaunitz gained information from Vienna's black chamber.

CHAPTER 3
HOW TECHNOLOGY CHANGED CRYPTOLOGY

In 1844, the electric telegraph changed communication and cryptology. Telegraph wires sent electric signals over long distances almost instantaneously. People created messages by tapping a switch on the telegraph. The taps sent electric currents through the wires. The length and pattern of the taps represented letters.

This language was called Morse Code. Morse Code was a public language, rather than a secret. For the first time, cryptology was the key to everyday communication among citizens.

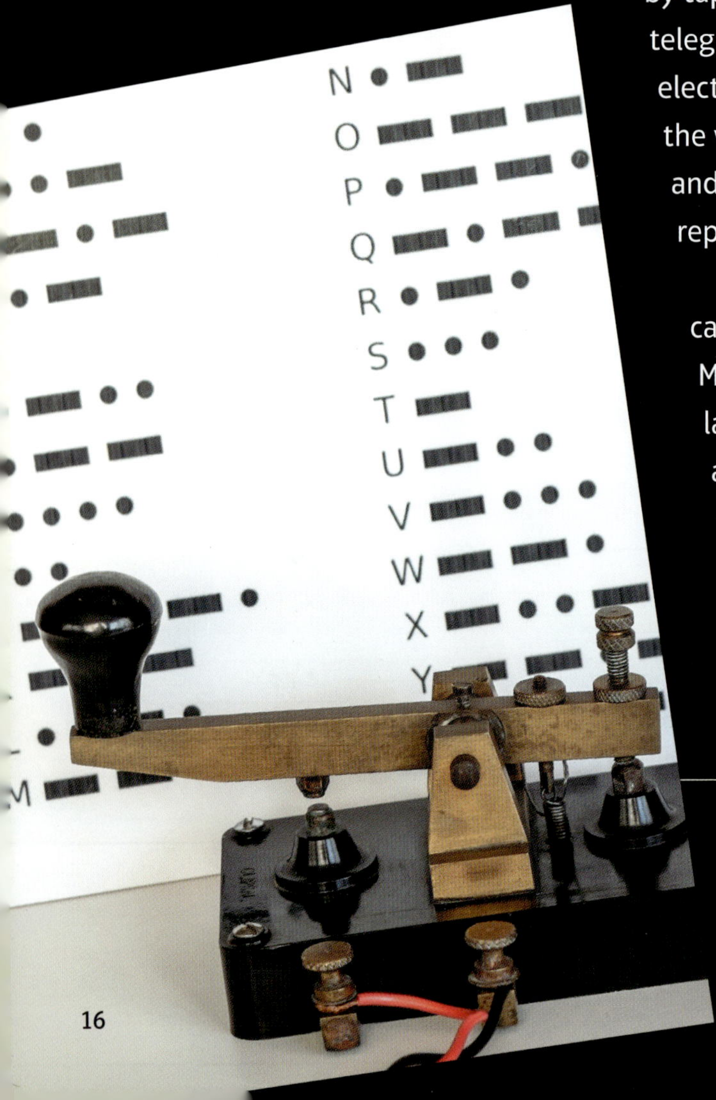

An electric telegraph in front of a Morse Code key